HEINEMANN STATE STUDIES

D1490001

Virgin

Plants and Animals

Karla Smith

Heinemann Library
Chicago, Illinois

Customer Service 888-454-2279

Visit our website at www.heinemannlibrary.com

Designed by Heinemann Library
Printed and bound in the United States by
Lake Book Manufacturing, Inc.

07 06 05 04 03
10 9 8 7 6 5 4 3 2 1

Library of Congress
Cataloging-in-Publication Data

Smith, Karla, 1947-
Virginia plants and animals / Karla Smith.
p. cm.
Summary: Discusses the native plants, animals, and ecosystems of the
state of Virginia.
Includes bibliographical references and index.
ISBN 1-4034-0360-0 (hardcover) -- ISBN 1-4034-0582-4 (pbk.)
1. Natural history--Virginia--Juvenile literature. [1. Zoology--Virginia. 2. Botany--Virginia. 3. Biotic communities--Virginia.] I. Title.
QH105.V8 S65 2003
578'.09755--dc21

2002152967

Special Thanks

The author would like to thank naturalists Jean Hodges and Catherine Roberts for their expert advice on the content of this book.

The publisher would like to thank Gary Barr for his comments in the preparation of this book.

Some words are shown in bold, **like this.** You can find out what they mean by looking in the glossary.

Acknowledgments
The author and publishers are grateful to the following for permission to reproduce copyright material:

Cover photographs by (main) Leonard Lee Rue III; (row, L-R) Jerome Wexler/Photo Researchers, Inc., Tom Brakefield/Corbis, Alex Brandon/Heinemann Library, John D. Cunningham/Visuals Unlimited, Inc.

Title page (L-R) Nicholas Bergkessel, Jr./Photo Researchers, Inc., Dave B. Fleetham/Visuals Unlimited, Inc., John D. Cunningham/Visuals Unlimited, Inc.; contents page (L-R) David A. Northcott/Corbis, Bob Jensen/JENSE/Bruce Coleman Inc., Ron Austing/Frank Lane Picture Agency/Corbis; p. 4 Richard T. Nowitz/Corbis; p. 5T Jerome Wexler/Photo Researchers, Inc.; p. 5B Nancy Rotenberg/Earth Scenes; p. 6 Tom Brakefield/Corbis; pp. 7, 9, 18T David Muench/Corbis; p. 10 Ron Austing/Frank Lane Picture Agency/Corbis; p. 11 Richard Day/Animals Animals; pp. 12, 25, 29B, 31 Lynda Richardson/Corbis; p. 13 Frank Blackburn/Ecoscene/Corbis; pp. 14T, 28T, 30 Rob and Ann Simpson/Visuals Unlimited, Inc.; p. 14B Nicholas Bergkessel, Jr./Photo Researchers, Inc.; p. 15 Bob Jensen/JENSE/Bruce Coleman Inc.; p. 16 Raymond Gehman/Corbis; p. 18B John D. Cunningham/Visuals Unlimited, Inc.; p. 19 David A. Northcott/Corbis; p. 20T Doug Wechsler/Animals Animals; p. 20B John Bova/Photo Researchers, Inc.; p. 21T Bates Littlehales/Animals Animals; p. 21B Nik Wheeler/Corbis; p. 22T Leonard Lee Rue III; p. 22B Adam Jones/Visuals Unlimited, Inc.; p. 23T Dan Guravich/Corbis; p. 23B Swift/Vanuga Images/Corbis; p. 24T Dave B. Fleetham/Visuals Unlimited, Inc.; p. 24B Zig Leszczynski/Animals Animals; p. 26 W. Gregory Brown/Animals Animals; p. 28B Photo Researchers, Inc.; p. 29T Maresa Pryor/Animals Animals; p. 32 William and Mary College; pp. 33, 44B Bettmann/Corbis; p. 34 Academy of Natural Sciences of Philadelphia/Corbis; p. 35 Scott T. Smith/Corbis; p. 36 Mark Newman/Visuals Unlimited, Inc., Inc.; p. 38 Chris Mattison/Frank Lane Picture Agency/Corbis; p. 39 Dave G. Houser/Corbis; p. 40 Joe McDonald/Visuals Unlimited, Inc.; p. 41 Jamie Baxter/Save the Bay-Chesapeake Bay Foundation; p. 42 Buddy Mays/Corbis; p. 43 Joe McDonald/Corbis; p.44T Rob Simpson/Visuals Unlimited, Inc.

Photo research by Julie Laffin

Every effort has been made to contact copyright holders of any material reproduced in this book. Any omissions will be rectified in subsequent printings if notice is given to the publisher.

Contents

Wild Virginia

Virginia's location, land, and **climate** create many **ecosystems** and **habitats** for a variety of plants and animals. A habitat is the natural **environment** of a specific plant or animal. Habitats overlap in communities that depend on each other. These communities are called ecosystems. Forests, meadows, **wetlands,** coastlands, rivers, streams, lakes, and caves are ecosystems that have many habitats within them.

Many wild horses live in the coastal wetlands of the Chincoteague National Wildlife Refuge.

Virginia Ecosystems

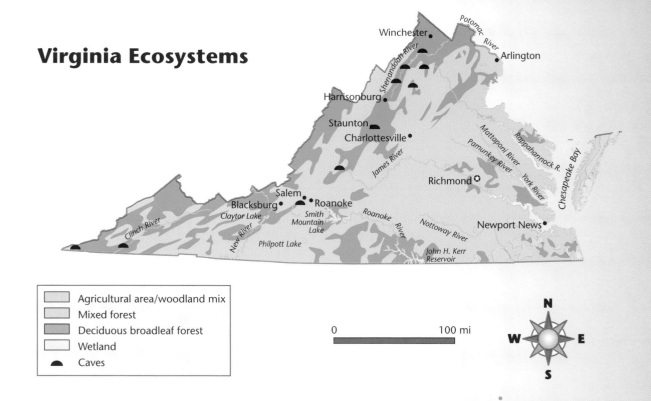

Legend:
- Agricultural area/woodland mix
- Mixed forest
- Deciduous broadleaf forest
- Wetland
- Caves

0   100 mi

Virginia has a large variety of plants and animals in its many different habitats and ecosystems.

DOGWOOD

In March of 1918, the flowering dogwood became the official state tree and flower of Virginia. Dogwood trees are small trees that grow everywhere in the state. The flowers on the dogwood tree bloom beginning in mid–April and fill the forests with their white blossoms. The dogwood also produces a red berry that stays on the tree well into the winter. The berries are a favorite food of **game** birds, songbirds, chipmunks, mice, and deer.

Virginia's flowering dogwood trees have white blossoms and red berries.

*The wild turkey is the largest **game** bird in North America.*

WILD TURKEY

Wild turkeys **forage,** or search, for nuts, seeds, acorns, and insects to eat. They live in open woodlands, forest clearings, and meadows. At night they settle down in oak or pine trees. They can be found all over Virginia, and usually travel in groups called flocks. Males grow to be about four feet long, including the tail, and females grow to be about three feet long.

Reporting Back to England

John Smith, one of Virginia's first English colonists, was amazed at the **natural resources** he found in Virginia in 1607:

There is excellent land full of flowers of different kinds and colors, and as goodly trees as I have seen, as cedar, cypress, beech, oak, walnut, sassafras, and vines in abundance, whose grapes hang in clusters to many trees. There are also many fruits . . . bigger and better than ours in England, mulberries, raspberries, and fruits unknown. In the rivers are great plenty of fish of all kinds and as for sturgeon, all the world cannot be compared to it. Also in this country are many great and fair meadows, low **marshes** having excellent pasture for cattle. There is also great store of deer, and wild animals as bears, foxes, otters, beavers, muskrats, and wild beasts unknown.

Forests

Forests cover more than half of Virginia. Forests are one of Virginia's most valuable natural resources. Plants, trees, and many kinds of animals make their homes in the forest **habitats.**

KINDS OF FORESTS

Virginia has an amazing variety of types of forests. Each kind creates habitats for a variety of plants and animals.

Maritime forests in Virginia are located along the edge of coastal beaches near the Atlantic Ocean and Chesapeake Bay. The plants there must be able to survive salt spray, flooding with ocean water, and high winds. Because of these conditions, pines and live oaks that grow in the maritime forests do not grow as tall as they do elsewhere and are bent with the wind. Maritime forests also include plants such as wax myrtle, American holly, and yellow jasmine.

*In Seashore State Park in Virginia Beach, a maritime forest grows beyond the shore of a tidal **marsh.***

Virginia Food Web

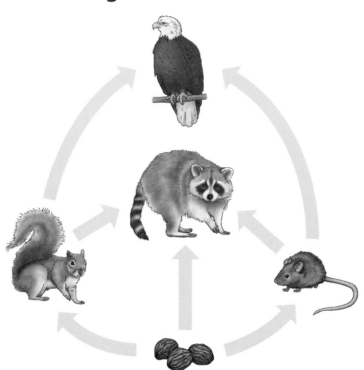

*Every habitat in Virginia has a specific food web. If any plant or animal within that food web should die out, the others could become **threatened** as well.*

Many of the animals found in the maritime forests are just passing through. Deer, raccoon, mice, and many kinds of birds feed on nuts, berries, and plants in these forests. The barred owl and the fox hunt small animals in this **habitat.** Insects such as mosquitoes, horseflies, and ants are also part of this **food web.**

Pine and hardwood mixed forests—often called temperate forests—are located in the Coastal Plains region, the Piedmont, and the mountains. Virginia was once filled with huge oak, hickory, chestnut, walnut, and poplar trees. Then, beginning in 1607, settlers cut down forests to make room for farms. Later, when some farm fields were abandoned, fast-growing pines, maples, and sweet gums were the first trees to grow there. Hardwood trees soon followed. Today, more than 70 percent of Virginia's forests are a mix of pine and hardwood trees.

Alpine evergreen forests grow at high elevations in very cold **climates.** In Virginia, these forests are found near the tops of mountains. Fraser fir and red spruce grow there. Plants such as rhodendrons, flame azaleas, and mountain laurel also grow there.

One of the last natural stands of the longleaf pine is located near Zuni, Virginia, and is a protected area.

LONGLEAF PINE

The longleaf pine is found only on the coastal plains of southeast Virginia. These trees were once plentiful in the state. They were valuable in colonial days for shipbuilding. The tall, straight trunks were used for making sailing ships.

Because so many longleaf pines were cut down, very few of these kinds of pines are found in Virginia today. But several organizations in Virginia are working to help restore and preserve the longleaf pine.

WHITE OAK

The white oak tree is one of Virginia's most important trees for lumber. The fruit of the white oak is an acorn. Squirrels collect, save, hide, and feast on them. Like the longleaf pine, the white oak was once used as timber for making ships. Today, the wood is used to make furniture and houses.

SASSAFRAS

The small sassafras tree—part of the pine hardwood mixed forests of Virginia—was one of the first **natural resources** sent to England from Jamestown in 1607. The English used the sweet smelling oil of sassafras for medicine and flavoring. Later, sassafras root was used to make tea and root beer, and for flavoring.

SPRUCE AND FRASER FIRS

Alpine forests grow in the upper elevations, in Virginia's mountain areas. Trees such as the red spruce and the fraser fir grow well in the high elevations of Mount Rogers, the highest point in Virginia.

BIRDS OF THE FOREST

Birds depend on the mixed hardwood forest for their **habitat.** The trees and flowers provide seeds and berries for food. Many **species** of birds build their nests in the tree branches. Songbirds such as cardinals and wrens build nests in shrubs closer to the ground. Other birds, such as owls, make homes in trunk hollows.

The pileated woodpecker can grow to be sixteen inches long.

PILEATED WOODPECKER

The pileated woodpecker found in Virginia's forests is the largest species of woodpecker in North America. Woodpeckers eat insect **larvae** that burrow in tree bark. The woodpecker pecks the bark away and then reaches its sticky tongue as far as three inches beyond its bill to snag its **prey.**

CARDINAL

The cardinal became the state bird of Virginia in 1950. The cardinal does not **migrate**—it lives in Virginia year round. It is a bright red bird. The male has a crest of feathers on its head and a black patch around a triangular red bill. The female is brownish with a crest and black patch around the bill. These dull colors hide her when she sits on a nest, which helps protect the nest and eggs from **predators.**

MAMMALS

Small **mammals** like mice and chipmunks live on the forest floor. Squirrels, opossums, and bats spend much of their time in trees. Opossums eat small **rodents,** snakes, frogs, birds, bird eggs, wild nuts, and fruits.

Squirrels are plentiful in Virginia, too. The gray squirrel is the most common and lives in oak and hickory trees. They build leafy nests in the treetops. Squirrels feed on nuts, seeds, fruits, and spring tree flowers.

Larger forest mammals such as deer, bears, and bobcats **forage** and hunt for their food.

BLACK BEAR

The black bear is the largest mammal in Virginia. These bears live in the state's mountain forests. They feed on insects, frogs, fish, and nuts. Bears also like blueberries, strawberries, and apples. Black bears sleep in a protected place, such as an old log, for part of the winter.

WHITE-TAILED DEER

Virginia's white-tailed deer is the state's most important big-game animal. It is called the white-tailed deer because the underside of its tail is white.

In 1937, there were only two black bears reported living in the Virginia mountains. Because of new hunting laws, there are now between 300 and 800. The black bear is the only wild bear that lives in Virginia. Many live in Shenandoah National Park and the Great Dismal Swamp.

The deer's favorite **habitat** is the edge of the forest, near farm areas and close to water. Deer spend the day **camouflaged** in the forest. They feed at dusk or dawn on grasses and water plants.

OTHER ANIMALS

Many **amphibians** and reptiles make their homes on the forest floor. They hunt for food among the leaves and hide from **predators.** One kind of amphibian, the pigmy salamander, is the smallest salamander in Virginia. It is found only in forests in the mountains above 4,000 feet.

There are 30 **species** of snakes in Virginia. Water snakes such as the northern water snake and the ribbon snake are very common. Most land snakes are camouflaged to blend in with their habitat. Box turtles and blue-tailed skinks are other kinds of reptiles that make their homes in the forests.

Insects are an important part of the **food web** in the forest. Ants and beetles are **scavengers** among the leaves. Mosquitoes, flies, and ticks get their food by sucking blood from **mammals** and birds. The insects then become meals for the birds, bats, and snakes living in the trees. There are more than 20,000 species of insects in Virginia.

Virginia's five-lined skink has a bright blue tail that can be spotted easily.

Meadows

Most meadows are located on the edges of forests. These open fields occur when the forest is cleared by natural fires or by humans for farming. Bushes, wildflowers, and grasses grow in meadows. Blackberries, wild strawberries, raspberries, and blueberries also grow in open meadows. They were an important food source for Native Americans and colonists. Berry bushes make good habitats for small animals.

Cold temperatures at high elevations create arctic meadows. Mount Rogers, Buffalo, Hawksbill, and Stoneyman Mountains have plants that cannot be found anywhere else in Virginia.

MEADOW PLANTS

The open tops of Virginia's mountains have small arctic plants, such as mountain sandwort and Indian paintbrush. Thornless blackberry bushes and rosebay grow below the tops of these high peaks.

Rosebay willowherb, or fireweed, is a common woodland meadow weed that grows all year.

Jimsonweed is a tall plant with a purplish or greenish stem. It has a trumpet-shaped white or violet flower. All parts of jimsonweed are poisonous.

March, April, May, and June are good months to find wildflowers blooming throughout Virginia. Wildflowers bloom in June and July in high meadows in the Appalachian Mountains. There are more than 900 different kinds of wildflowers in the mountains alone.

Jimsonweed, or Jamestown weed, can be found throughout Virginia. The name *jimson* came from the word *Jamestown,* because jimsonweed grew near the colonists' homes.

MEADOW ANIMALS

Birds, small **mammals,** and insects live in meadow **habitats.** The **endangered** snowshoe hare and northern flying squirrel live in Virginia's arctic meadows.

The swallow family of songbirds eats insects. Purple martins are the most common of the swallow family. Each spring they arrive and nest in trees at the edge of meadows. They help people by eating insects. Purple martins can eat up to 2,000 mosquitoes a day!

*Trumpet honeysuckle is the only honeysuckle plant **native** to Virginia. It is also called coral honeysuckle. It is a vine that climbs fences and other plants to reach sunlight. The trumpet honeysuckle is a favorite food of the hummingbird.*

The smallest bird found in Virginia is the ruby-throated hummingbird. It is a **migratory** bird that spends summers in North America and winters in South America. The hummingbird depends on nectar from flowers for its food.

Hawks and owls search the meadows for **prey.** Their sharp eyesight helps them to find small mammals and birds **camouflaged** in the underbrush.

Most of Virginia's **game** birds, such as the pheasant, bobwhite quail, and ruffed grouse, feed on the insects, leaves, and berries in the meadows. They build their nests on the ground among the plants.

Small mammals like field mice and rabbits survive on seeds, roots, and fruit from meadow plants. Red foxes and striped skunks are **nocturnal** animals that hunt smaller mammals at night. Moles burrow underground, where they feed on worms and are safe from the sharp eyesight of **predators.** Bears and deer also visit meadows to feed.

Insects are plentiful in meadows, too. Some insects suck juices from plant stems or drink the nectar from flowers, while other insects feed on the blood of animals. Insects found in the meadows include grasshoppers, mosquitoes, ladybugs, beetles, dragonflies, ticks, bees, and wasps. Butterflies are some of the most brightly colored insects in a meadow.

The tiger swallowtail that flutters through Virginia's meadows was adopted as the state insect in 1991.

Wetlands

Virginia has about one million acres of **wetlands.** These are places that are covered with shallow water much of the year or have water-logged soil. Wetlands provide a unique kind of **ecosystem.** The Tidewater, or Coastal Plains region, in Virginia is part of a large wetland area.

There are two basic kinds of wetlands: coastal and inland. Coastal wetlands are affected by tides. They may be covered with water at one time of the day and dry at other times. Plants and animals that live in these wetlands have **adapted** to these changing conditions.

Brackish marshes are coastal wetlands in places where freshwater and saltwater mix, such as the lands along the rivers and **tributaries** of the Chesapeake Bay. Salt marshes are generally the kind of wetlands found along the beaches of Virginia's Atlantic coast.

Flocks of geese are often seen flying over Chesapeake Bay wetland areas.

Inland weltands, such as swamps, are most often fresh-water wetlands. They are not affected by the tides.

CHESAPEAKE BAY AND TIDAL RIVERS

The Chesapeake Bay is an ecosystem that covers 2,500 square miles. The southern half of the bay is within Virginia's boundaries. It is the largest **estuary** in the United States. The rivers that empty into the Chesapeake Bay are brackish water systems. These rivers get their freshwater from the mountain stream **runoff.** After the rivers pass the **fall line,** saltwater brought by the tide mixes with the freshwater.

PLANTS OF THE BRACKISH MARSHES

A unique world of plants lives in the Chesapeake Bay and its tributarics. The plants are known as **SAVs,** which stands for "submerged aquatic vegetation." They help create the nurseries for baby fish, oysters, crabs, and clams. Most of these plants have roots and grow entirely underwater. Wild celery, waterweed, and

Submerged Aquatic Vegetation (SAV) Restoration

Underwater grasses are important in the wetlands of the Chesapeake Bay. They help hold soil in place and are good hiding places for young fish and crabs. Many plants have been uprooted by boat activity. Freshwater from hurricanes has killed off many other plants. Students are helping to restore the underwater **environment** by growing eel grass and wild celery in containers of brackish water in the classroom. They control the level of salt, temperature, and light to get the best growing conditions. In May, the students plant the grasses in areas that need it.

Cordgrass has strong roots and holds its place along shorelines, helping prevent **erosion.**

pondweeds are some common SAVs. Small fish feed on **plankton** that gets caught in the grasses and their roots.

Other kinds of plants also live in the **brackish marshes** of the Chesapeake Bay. They include cordgrass, narrow-lined cattail, and black needlerush.

The blue crab is one of Virginia's most recognizable **crustaceans.** *It is dark green on top of its shell, white underneath, and has blue claws.*

SHELLFISH

Many kinds of animals depend on the **wetlands** in the Chesapeake Bay. Most of the animals found in **estuaries** live underwater. Crabs, oysters, clams, **mussels,** and other shellfish live on the bottom. Above them swim many kinds of fish that are found in the open ocean **habitat.**

Oysters are **mollusks.** They have a soft body that is protected by a hard shell. Oysters are considered the heart of the Chesapeake Bay's web of life, because they are excellent water **filters.** A single adult oyster can filter 60 gallons of water a day. The oyster removes plankton and **algae** from the water for food. It also removes **sediments.**

FISH

Many fish live in Virginia waters in the ocean and near the mouth of the Chesapeake Bay, where the water is very salty. These include striped bass, menhaden, and flounder. Some of these fish can be found in tidal waters in summer and fall. They **migrate** to warmer waters in winter.

SHAD

Shad are **anadromous** fish. This means they are born in rivers and then migrate to the ocean, where they spend most of their lives. Their average life span is five years. In those five years, a shad might migrate a total of 12,000 miles. When they are about five years old, shad travel to their home rivers to lay eggs. Baby shad spend their first summer in the freshwater streams. Cooler water temperatures in fall signal to the fish that it is time to migrate to the ocean.

REPTILES AND AMPHIBIANS

Brackish marshes are home to many kinds of reptiles and **amphibians.** The most common marsh reptiles are snakes and turtles. The copperhead, a poisonous snake, can be found in the wetlands of the Chesapeake Bay. Frogs, toads, salamanders, and newts are some of the kinds of amphibians that live in that habitat.

The diamondback terrapin feeds on oysters, crabs, and clams in Virginia's marshes.

Canada geese eat plants along the shores of tidal rivers.

MIGRATING BIRDS AND THE ATLANTIC FLYWAY

The honking of Canada geese was a signal for the Native Americans of Virginia that fall and winter were coming. It is still a sign that fall has arrived in Virginia's coastal wetlands. Swans, geese, ducks, and other birds travel thousands of miles from arctic nesting grounds to warmer places in winter. The path they travel is called the Atlantic Flyway. Many flocks of Canada geese have become permanent **residents** in areas where they can find food year round.

Seaside goldenrod has a stem with clusters of golden yellow flowers. Sparrows eat its seeds.

BALD EAGLES

The Chesapeake Bay has many bald eagles. The best eagle **habitats** are shoreline forests near plentiful fish. Bald eagle nests are usually built of large sticks and lined with soft materials such as pine needles and grasses. An eagle has sharp eyesight and soars high above its hunting area, waiting to spot **prey.**

SALT MARSH PLANTS

Salt **marsh** plants include cattails, bulrushes, cordgrasses, and saltgrass. The grasses have hollow stems and grain-like seeds. Higher on the banks of the marsh are plants with flowers, such as the seaside goldenrod.

SALT MARSH ANIMALS

Muskrats live anywhere there is water. Muskrats make burrows high on the banks of rivers, streams, and marshes. They eat water lily roots, cattails, and other plants. They also eat shellfish, crawfish, and turtles.

Herons and egrets are long-legged wading birds. They hunt fish, frogs, crabs, and **mussels** along the edges of salt marshes. They feel around with their feet and catch their dinner. Herons use their spear-like bills to open the shells of mussels.

SWAMPS

Freshwater swamps are habitats in which plants grow in wet soil for all or most of the year. Swamps are forested **wetlands.** The largest swamp in Virginia is the Great Dismal Swamp. Before colonists arrived in 1607, bald cypress, swamp black gum, and tupelo trees grew in the deepwater areas along the shores of Lake Drummond. Atlantic white cedar, or juniper, also grew in the wet soil. Colonists used its wood for roof shingles and barrel making. In the last 400 years, most of the Atlantic white cedar has been logged from the swamp. Red maple, black gum, swamp chestnut, oak, and swamp magnolia have grown in its place.

SWAMP PLANTS

The bald cypress is found only in swampy areas in Virginia. Its root tips grow above water to get oxygen, and are called cypress knees. Many of the bald cypress trees found in First Landing Park along the coast are covered with Spanish moss.

The great blue heron is the largest salt marsh heron.

Cypress trees grow throughout Virginia's Great Dismal Swamp.

The river otter can grow up to three feet long.

SWAMP ANIMALS

The Dismal Swamp is the **habitat** of several **mammals** that are **threatened,** including the southeastern shrew, eastern big-eared bat, river otter, and bobcat. Black bears and canebrake rattlesnakes also live in the swamp.

The river otter is a mammal that lives in unpolluted streams, ponds, and swamps. The otter is a fast swimmer, but can also travel on land. Otters eat fish, crabs, clams, oysters, and crawfish. They make their homes in old beaver lodges or in empty muskrat burrows.

Bobcats are medium-sized cats with very short tails, pointed ears, and long legs. Bobcats are grayish brown with black spots and bands. Bobcats have been sighted in the mountains and in the Dismal Swamp regions. They eat rabbits, squirrels, mice, and muskrats. They also eat turkeys and small deer.

Bobcats are protected from being hunted in the Great Dismal Swamp National Wildlife Refuge and in national parks.

Coast

Sandy beaches and dune habitats are found where the ocean meets the land. Sometimes the tide washes over them. Beach plants, such as American beachgrass, stand up well under these harsh conditions.

Beaches provide a feeding area for thousands of shorebirds. Birds such as gulls and terns survive by feeding on insects, fish, and **crustaceans.** These shorebirds are food for other animals, such as the peregrine falcon. Dunes provide nesting places for birds such as common terns and black skimmers. Crabs, clams, fiddler crabs, beach fleas, and insects are some of the other kinds of animals that live in this harsh and salty habitat.

Groups of black skimmers nest on Cedar Island.

OPEN OCEAN

Virginia has more than 200 miles of coastline bordering the Atlantic Ocean. The open ocean **ecosystem** touches its beaches. Dolphins, humpback whales, and fin whales can be spotted swimming along the coast throughout the year.

OCEAN MAMMALS

Humpback and fin whales are baleen whales. These whales feed by swallowing and forcing

Bottlenose dolphins play in the waters off Virginia's coastline.

Sharks are at the top of the food chain in Virginia's coastal waters.

the seawater through the baleen plates in their mouths. The baleen collects the **plankton** and small fish and traps it in the whale's mouth.

Dolphins and porpoises are small, toothed whales. They stay year round near the mouth of the Chesapeake Bay hunting for fish to eat. They often swim into the harbor following schools of fish. The most common dolphin seen in Virginia's coastal waters is the bottle-nosed dolphin.

SHARKS, RAYS, AND OTHER FISH

Between 19 and 35 different kinds of sharks swim in Virginia's coastal waters during summer and fall. The sharks are **migrating** in search of food at those times. The most common sharks in the area are sand tiger, dusky, and bull sharks.

The bat ray is also called a cownose ray because of its shape.

There are several kinds of rays swimming in the Chesapeake Bay and open ocean. They include the southern stingray, Atlantic stingray, bluntnose stingray, and bat ray. They feed on **mollusks,** such as clams. The base of the ray's tail contains a poisonous spine. Rays look like they have wings as they swim through the water.

John Smith and Stingray Point

Captain John Smith, a colonial explorer, learned about the Atlantic stingray the hard way. In the summer of 1608, Smith and a group of fourteen men explored the Chesapeake Bay, looking for a passage to the "East China Sea." They explored as far north as Baltimore. On their return trip to Jamestown, they stopped at the mouth of the Rappahannock River. Smith was stung in the hand by a stingray. John Smith's entire arm swelled and his men thought he would die, so they returned quickly to Jamestown. Smith survived, but that point of land where he was stung is known to this day as Stingray Point.

Virginia is the part-time home to many fish that **migrate** to warmer waters in the winter. Shad, striped bass, and sea trout are just a few of the 286 **species** of fish that can be found in Virginia waters.

SEA TURTLES

The coastal ocean waters near Virginia and the Chesapeake Bay are home to sea turtles. These turtles cannot pull their heads or flippers into their shells. Loggerheads and other sea turtles visit the Chesapeake Bay in summer to **forage** for food.

*Loggerhead sea turtle hatchlings have a dangerous walk from their eggs to the safety of the water. **Predators** snatch up as many of them as they can.*

Freshwater Streams, Rivers, and Lakes

Rivers and streams west of the **fall line** are freshwater **habitats** for many animals. Virginia has only two natural lakes. Its largest natural lake is Lake Drummond, in the Dismal Swamp. The second is Mountain Lake, near Blacksburg. In the 1900s, dams were built on several rivers in the Piedmont and Ridge and Valley regions. These dams created large man-made lakes. Virginia has several important rivers, many of which flow from west to east and empty into Chesapeake Bay.

Trout eat smaller fish, insects, crayfish, salamanders, and frogs.

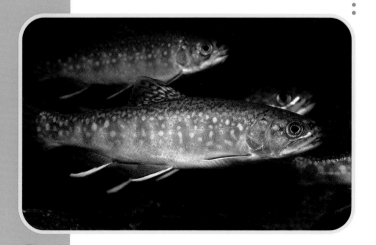

Fish

Virginia has more than 210 **species** of freshwater fish living in its rivers and lakes. Sunfish, striped bass, perch, trout, and catfish are some of the favorites of sport fishers. Small mouth and large mouth bass are popular, too. One of the best small mouth bass fishing spots in the country is on the James River in downtown Richmond.

Rivers and Lakes in Virginia

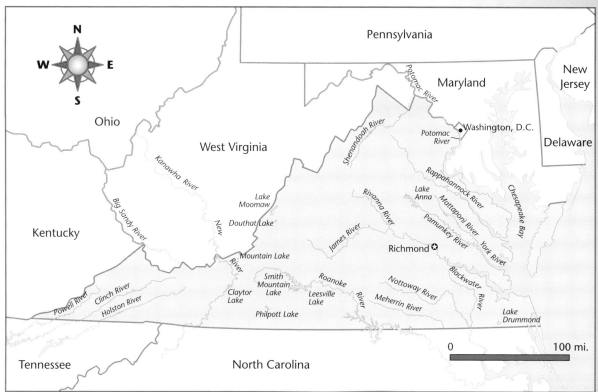

At one time, the brook trout was Virginia's only trout species. Then rainbow trout were introduced from the Pacific Coast, and brown trout were brought in from Europe and western Asia. They were put in coldwater creeks and rivers in the western part of the state.

*The freshwater provided by Virginia's many rivers is one of the state's most valuable **natural resources.** The rivers also are habitats for plants and animals.*

MUSSELS AND SNAILS

Aquatic mussels and snails live in freshwater streams and rivers. They are part of a **food web** that includes **algae,** decaying leaves, plants, insects, and small fish.

Mussels are **filter** feeders and help keep the rivers and streams clean. Mussels serve as food for muskrats, otters, and fishes.

Beavers are known for their building skills.

BEAVERS

Beavers live along freshwater streams and ponds. A beaver's favorite food is green tree bark. Beavers cut poplars and other trees along the edges of water with their sharp teeth. The top of the tree then falls over, and the beavers have dinner. Beavers dam up streams to make ponds for their houses. Their lodges are made with tree branches, mud, and stones. Beavers have silky-smooth fur that was valuable to Native Americans and colonial trappers. The beaver was almost hunted to **extinction** in North America in the 1800s. Laws regulating the hunting of beaver have helped to keep it from becoming extinct.

TURTLES AND LIZARDS

There are 22 kinds of turtles in Virginia. Freshwater turtles live in ponds, lakes, and along rivers and streams. Turtles eat crayfish, tadpoles, and **aquatic** plants. Raccoons, skunks, opossums, and some wading birds often eat turtles.

There are nine kinds of lizards in Virginia. Lizards feed on spiders, crickets, cockroaches, beetles, snails, other lizards, and small snakes. The eastern glass lizard is a

The eastern glass lizard is often mistaken for a snake, because it has no legs.

rare lizard that lives in False Cape State Park in the southeastern corner of the state.

SALAMANDERS

Virginia's rivers and lakes are home to 73 **species** of **amphibians.** Amphibians are animals that live on both water and land during their life. Salamanders, frogs, and toads are in this group. There are 47 species of salamanders found in Virginia. Salamanders depend on forests and good clean water to survive. They are amphibians with smooth moist skins and toes without claws.

Hellbenders can grow to two feet or more in length and are **aquatic**. *Some hellbenders live to be 25 years old.*

The largest salamander found in Virginia is called the hellbender. Hellbenders live in clear freshwater streams that have plenty of flat rocks. They remain under cover by day and look for food at night on stream bottoms. They eat crayfish and fish. Hellbenders are found along the New, Holston, Clinch, and Powell River systems.

FROGS AND TOADS

There are 26 different kinds of frogs and toads found all over Virginia. Frogs and toads lay their eggs in water. They go through stages of change until they can be land animals. Most frogs and toads return to the same stream, pond, or pool to breed. The common toad has been called the "farmer's friend" because it helps farmers by eating insects in gardens and fields.

The American toad is found in all but the southeast of Virginia.

Caves

There are thousands of caves located in the Ridge and Valley region of Virginia. In fact, a survey taken in 2000 counted 3,641 caves in Virginia. The deepest known cave in Virginia is 1,260 feet deep. Caves are dark, wet, and cool **environments.**

Virginia caves vary in size. Some are only large enough to crawl into. Other caves are interconnected passages that extend for twenty miles.

Caves are a unique **habitat.** Unlike most other habitats, no plants grow in caves. This is because caves are places where the sun never shines. They are always in total darkness.

Caves are also places that keep a steady temperature. The caves in Virginia are usually between 54 and 56 degrees. Temperature changes due to day and night or changing seasons only affect the temperature at a cave's entrance. The temperature of the rest of the cave is the same during the day and night and throughout the year.

BATS

Bats are the most famous cave-dwelling animals. Thousands of bats **migrate** to Virginia to **hibernate** in caves from October until April. They cling upside

The silver-haired bat is medium sized with silver tips on its brownish-black hairs.

Life in the Dark

One kind of cave animal is found only in one cave system in Lee County, Virginia. It is an **isopod.** Over the years, isopods have adapted to their cave environment in such a way that they could no longer survive outside the cave. This is largely due to the fact that they are eyeless. Luckily, they do not need sight while in the total darkness of a cave.

down from the ceilings in huge clumps. Baby bats, called pups, are born in June and July. They feed on their mother's milk for only 25 to 35 days. Then their mothers teach them to fly and **forage** for food on their own.

Bats live in forests and near fields all over North America in the summer. They are **nocturnal** and fly through the night sky scooping up thousands of insects. Bats have built-in sonar that helps them find their way in the dark. Their hands resemble human hands with a thumb and four fingers. Their skinny fingers form the frame for their webbed wing.

OTHER ANIMALS

Virginia's caves are home to insects such as camel crickets and daddy-longlegs. Salamanders also use the caves as protection against **predators.** These animals, however, have to leave the cave to find food.

There are some animals, however, that have totally **adapted** to living in the cave. These animals never leave their cave.

Extinct Species

The land we call Virginia has changed many times since it formed. The Appalachian Mountains are some of the oldest mountains on Earth and were once as high as the Alps. **Erosion** slowly wore the mountains down and created the Piedmont and Coastal Plains regions.

Periods of cold and warm **climates** created **habitats** for different **species** of plants and animals. At times, the Coastal Plains region was under water. Other times, Virginia was covered with tropical forests. When the climates changed, some plants and animals **adapted** and survived. Other species could no longer survive in Virginia, but lived in other regions. Species that die out in one area but survive in another are called **extirpated.** Some species could not adapt and died out everywhere. They became **extinct.**

ANCIENT FOSSILS

When Earth was much warmer millions of years ago, Virginia was on the bottom of the ocean floor. Scallops, oysters, clams, and prehistoric fish lived in its waters. **Fossils** found along the coastal plains

Virginia's state fossil is the scallop Chesapecten jeffersonius.

tell us that water animals lived in the ancient seas in southeast Virginia before dinosaurs lived on Earth. When Earth turned colder, the sea level dropped. The animals that lived there lost their habitat, and many died and became extinct.

Fossils also tell us that tropical fern forests once covered Virginia, and dinosaurs roamed its hills. Dinosaur tracks have been found in Culpeper Stone Quarry and Pittsylvania County.

Fossils of the mammoth, mastodon, musk ox, and ground sloth were found near Saltville. These animals are no longer found in Virginia. Scientists are not sure whether the changing climate or hunting caused these animals to disappear. The mastodon, mammoth, and ground sloth are no longer found anywhere. They have become extinct.

The mastodon was an early relative of the modern-day elephant.

PEOPLE CHANGE THE ENVIRONMENT

Millions of years of natural climate and landscape changes could not match the impact people have had on Virginia's **environment.** Native Americans were careful about preserving the state's natural **habitats** and landforms. The European settlers who came were not as careful.

Many European settlers arrived and started farms in the wilderness. Trees were seen as raw materials to be cut and used, not as part of a balanced habitat. Wild animals

Chestnut Trees

The American chestnut was one of Virginia's most valuable trees, and it once grew all over the state. Squirrels, turkeys, bears, and people ate nuts from the chestnut tree, and people used the wood for building. A **fungus** invaded the bark of the tree in the early 1900s. This was called the chestnut blight. Virginia's chestnut trees were all dead by the early 1930s. The few full-grown American chestnut trees left are found in Michigan. Scientists hope to introduce a new variety of tree that can resist the chestnut blight.

were a source of food and were hunted for their furs. Colonists and early settlers were paid for killing wolves, because the wolves were a threat to farm animals.

When settlers first arrived in the Shenandoah area, **mammals** were plentiful. Explorers of the early 1700s mentioned seeing bison in the western parts of the Ridge and Valley region. Within a few years, bison, elk, panthers, timber wolves, black bears, and white-tailed deer were gone from the area. The Eastern bison, a small buffalo, is now **extinct.**

PASSENGER PIGEONS

In 1500, there were about 3 to 5 billion passenger pigeons in eastern North America. Colonists reported that the sky would turn black as

By 1914, passenger pigeons were extinct.

flocks flew by. These birds nested in forests in large colonies of thousands of birds. They ate acorns and beechnuts.

By the 1850s, however, many oak and beechnut forests were cut down. The **habitat** of the passenger pigeon was destroyed at the same time that millions of the birds were being killed for food.

RUNOFF

Freshwater streams in the western part of Virginia have hundreds of **species** of freshwater snails and **mussels,** both of which are **mollusks.** The **runoff** from coal mining and paper factories has changed the acid level in the river water. Mussels and snails are dying in these southwest rivers. The same runoff that has killed the mussels and snails also affects fish. Many kinds of fish are no longer found in Virginia's cold-water streams.

Living Fossil

The Atlantic horseshoe crab is a strange looking animal. It is the oldest **living fossil** found in Virginia. The horseshoe crab has been on Earth for about 360 million years. It belongs to the arthropod group of animals. They have five pairs of walking legs and ten eyes. Horseshoe crabs are more similar to spiders and ticks than they are to true crabs.

The horseshoe crab spawns and lays eggs on the Delaware and Chesapeake Bay beaches. It lays its eggs in nests in the sand between high and low tide. After about a month, the eggs hatch and swim in shallow pools until they mature.

Endangered Species

When people change the conditions in a **habitat,** many plants and animals cannot survive. Hunting and over collecting can also cause a **species** to become **extinct** or **endangered.** The U.S. Fish and Wildlife Service has 28 species of plants and animals on its endangered list for Virginia. The state of Virginia has its own list of endangered species that includes many other **threatened** species.

Mountain lions can live in mountains, woodlands, swamps, or caves.

MOUNTAIN LION

English colonists found mountain lions in the Coastal Plains region in the early 1600s. As settlers moved west, they saw more mountain lions. Mountain lions did not stay near human settlements. They fed on deer, elk, rabbits, squirrels, and smaller animals. In the 1800s, people hunted mountain lions until they were almost extinct. Only a few mountain lions have been spotted in the western mountains of Virginia in recent years.

SNOWSHOE HARE

The snowshoe hare is the largest of its kind. It changes color with the season. The snowshoe hare is losing its habitat. It can be found only in areas where red spruce, rhododendron, and mountain laurel grow. Increased deer populations have cleared out the underbrush that the hare depends on for food and protection.

U.S. Endangered Species in Virginia

Plants

sensitive joint-vetch

smooth coneflower

swamp pink

small whorled pogonia

Virginia spiraea

Animals

Indiana bat

Virginia big-eared bat

gray bat

Dismal Swamp southeastern shrew

Delmarva Peninsula fox squirrel

American peregrine falcon

red-cockaded woodpecker

Kemp's ridley sea turtle

leatherback sea turtle

loggerhead sea turtle

hawksbill sea turtle

slender chub

spotfin chub

duskytail darter

yellowfin madtom

fanshell

dwarf wedge mussel

cumberland monkeyface mussel

shiny pigtoe

BATS

The gray bat and the western big-eared bat are endangered. They **hibernate** in caves during the winter. When people explore caves, they disturb the hibernating bats. Scientists believe that even one person entering a cave can cause thousands of bats to wake up, thereby causing the bats to waste 10 to 30 days' worth of stored fat. Their fat supply must last until spring or the bats will starve. Other endangered bats in Virginia are the little brown bat and the eastern big-eared bat.

SEA TURTLES

The loggerhead is one of five kinds of sea turtles that visit Virginia waters. The habitat of the sea turtles is the open ocean and the Chesapeake Bay. Sea turtles **migrate** along the Atlantic coast when seasons change. Loggerheads are the only sea turtles that lay their eggs in the sands of Virginia's beaches. Beach development

The canebrake rattlesnake is a type of timber rattlesnake found in southeastern Virginia.

and vehicles on the beach have destroyed nesting grounds. Baby turtles hatch and crawl into the ocean. If the babies survive their crawl to the water and grow to become adult turtles, they must then survive commercial fishing nets and boat propellers. Other **endangered** sea turtles that **migrate** to Virginia waters include the Atlantic green sea turtle and the leatherback sea turtle.

CANEBRAKE RATTLESNAKE

Areas around the Dismal Swamp are being developed for houses and highways. The canebrake rattlesnake lives in and near the Great Dismal Swamp and near Northwest River Park in the city of Chesapeake. The canebrake rattlesnake was added to the endangered list because its **habitat** is being destroyed by development.

ENDANGERED FRESHWATER FISH

Most of the endangered freshwater fish in Virginia live in the southwest river systems of the Clinch, Powell, Holstein, and Big Sandy Rivers. Water and air pollution have changed the quality of water in these rivers. The disappearance of fish and **mollusks** from Virginia rivers is a warning that the rivers have a pollution problem.

NORTHEASTERN BEACH TIGER BEETLE

The northeastern beach tiger beetle has stripes like a tiger. It can only be found in a few sandy beach habitats along the Chesapeake Bay. The beetles lay their eggs in the sandy dunes. As more and more people build houses along beaches and drive vehicles on the dunes, the habitats of the beetle are destroyed.

PLANTS

There are many endangered plants in Virginia. When people build structures in natural areas, they often remove plants. Swamps and **wetlands** have been drained and forests cut down. The building of cities has changed plant habitats in the Coastal Plains region and at the **fall line.** Farming and lumbering have affected plant habitats in the Piedmont and the mountains.

SHADBUSH

Shadbush is an endangered shrub, or small tree, with white or pale pink flowers. It blooms from March to April. The name *shadbush* comes from Virginia's **native** peoples. They set up fish nets to catch shad each year when this plant bloomed. Shad migrate up the rivers in March to lay their eggs.

Fraser Firs

Fraser fir trees are almost **extinct** in Virginia. The woolly adelgid, an insect pest, has been killing them for the last 20 to 30 years. Woolly adelgids are insects that invade and destroy stems and branches of fir and balsam trees. Many eastern forests are experiencing this pest invasion.

Human Impact

Virginia's plants and animals are in a delicate balance. **Urban** areas have replaced many forest **habitats** of plants and animals. Animals that have **adapted** to these new places are known as urban wildlife.

URBAN WILDLIFE

Urban wildlife includes wild animals—not pets or farm animals—that live in cities and **suburban** areas. Most of these animals are birds or **mammals.** They live in cities and parks because human beings have created habitats there in which they can survive. Most of the urban wildlife **species** are plant-eaters. Backyard birdfeeders, artificial lakes, ponds, and even golf courses create an **environment** that provides food for wildlife.

Many **migrating** birds have become **residents** of Virginia's cities because the forests have been removed. They live in the environment created by parks and neighborhood landscaping. Many geese that usually migrate to and from Virginia in different seasons will

Wild skunks have adapted to life in urban Virginia, often searching through garbage cans for food.

Student BaySavers—Saving the Shad

Dams built on some of Virginia's rivers stopped many fish from getting to their home rivers to breed, and shad numbers declined. Passageways and fish ladders were constructed to help the fish return to their breeding grounds. By the time this was done, however, the shad had lost their instinct for finding their home rivers. Biologists from the U.S. Fish and Wildlife Service began to grow baby shad to release in cold-water streams. Now shad numbers are growing again in the waters of Virginia.

Students are helping to restore shad to Virginia's rivers. They get fish eggs from the U.S. Fish and Wildlife Service in April and raise the hatching fish in classroom aquariums. They feed the baby fish brine shrimp until they are old enough to be released. These fish instinctively return each year to the river where they were released.

remain year round if there is a food supply. When people feed Canada geese, the geese do not leave. Large flocks of Canada geese sometimes live in suburban neighborhoods and city parks and can become pests.

When animals find a constant supply of food, they stay. Raccoons, squirrels, rabbits, and even beavers have adapted to living in urban areas. Raccoons have learned to get into trashcans, and rabbits invade backyard gardens. The red fox and coyote look for food in urban neighborhoods when forests are cut down. When these animals adapt to life in the neighborhood, their **prey** becomes **rodents,** rabbits, and even small pets.

Black bears from the Dismal Swamp in southeastern Virginia frequently visit farms and neighborhoods looking for food. Urban growth has cleared much forestland, and the bear's natural habitat is shrinking.

Kudzu vines cover everything in their path, even tree trunks.

The smallest type of **urban** wildlife is the insect. Backyard gardens and parks need insects to **pollinate** plants and spread seeds. Small animals and birds eat insects. However, human beings and insects often clash. Many people consider mosquitoes, bees, and biting flies pests because they can all sting. Carpenter ants and carpenter bees can eat and bore their way through solid wood and destroy wooden buildings.

TRANSPLANTED PLANTS AND ANIMALS

European settlers brought seeds for wheat, vegetables, flowers, fruits, and herbs to grow on their farms in Virginia. Some of these things were harmful to the local **environment.** Settlers brought farm animals, too. Other animals have arrived by accident and have become **predators** to **native species.**

Dandelions, Japanese honeysuckle, and kudzu are all plants **transplanted** to Virginia. Japanese honeysuckle and kudzu are vines that can take over an area and choke out other plant life. The kudzu is slowly spreading along highways in Virginia.

The Norway rat and house mouse came to Virginia on ships with the early colonists. The Norway rat survives well in the wild and in the city. It likes to live under things and will eat anything that people eat.

REINTRODUCED SPECIES

The white-tailed deer and river otter were hunted out of the western part of Virginia. All large **mammals** were gone from the Blue Ridge region by the late 1800s. In 1934, the National Park Service imported thirteen deer into the area. Today there are about 6,000 deer.

Black bears returned to Virginia as the Blue Ridge and Shenandoah areas were allowed to return to forest. Bobcats, turkeys, and beavers have also returned.

WILDLIFE AWARENESS

Several organizations in Virginia are working to preserve wildlife habitats in the state. For example, the Virginia Department of Transportation (VDOT) helps to set up peregrine falcon nesting locations on state bridges. VDOT's Virginia Wildflower Program organizes the planting of wildflowers along state highways. The wildflowers are food for songbirds and beneficial insects. Garden clubs, civic organizations, and individual citizens all help by planting about 2,500 pounds of wildflower seeds each year.

Gypsy Moth

A small but dangerous transplant to Virginia is the gypsy moth. The gypsy moth was first found in Shenandoah National Park in 1969. It originally came from Europe to the United States as an experiment with silk worms.

The gypsy moth caterpillar eats leaves. In 1992, hatching caterpillars ate the leaves off the trees in almost 800,000 acres of forest in Virginia! That is an area over 35 miles long and wide. Gypsy moth caterpillars prefer oak trees and are ruining many trees in Virginia. Scientists are working hard to control and kill the caterpillars.

Virginia's wildflowers, such as the sea lavender, seaside gerardia, and glasswort shown here, need to be protected.

The Nature Conservancy is a national group with more than 36,000 Virginians working to preserve plants, animals, and natural communities. The Nature Conservancy owns and manages 32 preserves in Virginia. It has helped protect 225,000 acres of wildlife **habitat.**

The variety of plants and wildlife in Virginia is amazing. Virginia's landscapes, **climate,** and location create more **ecosystems** than most other states. Its wildlife is as plentiful as anywhere else in North America.

Peregrine Falcon

The peregrine falcon was **extirpated** from Virginia by 1960. Scientists think that **pesticides** wiped out the falcon in Virginia. DDT was used on farmlands to kill mosquitoes. Birds and **rodents** ate seeds and insects that had absorbed DDT. Falcons fed on these smaller animals. The DDT caused the falcon's eggshells to be too soft, and eggs did not hatch.

The peregrine falcon's natural habitat had been the cliff faces of the Appalachian Mountains, where it hunted rock doves. In the 1980s, peregrine falcons were **reintroduced** to Virginia. Scientists placed nesting boxes on bridges and tall buildings for the falcons. They have been tracking the falcon chicks as they leave the nest. Several pairs of peregrine falcons are beginning to reproduce. Now the falcons hunt pigeons in cities and birds and waterfowl near the bridge nests. The peregrines capture **prey** by hitting it in the air at dives of more than 80 miles per hour.

Map of Virginia

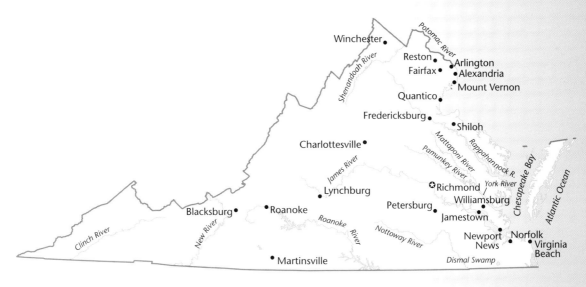

Winchester

Shenandoah River

Potomac River

Reston
Arlington
Fairfax
Alexandria
Mount Vernon

Quantico

Fredericksburg

Shiloh

Charlottesville

James River

Mattaponi River

Rappahannock R.

Pamunkey River

Chesapeake Bay

Atlantic Ocean

Lynchburg

Richmond

York River

Williamsburg

Blacksburg

Roanoke

Petersburg

Jamestown

New River

Roanoke River

Nottoway River

Newport News

Norfolk

Virginia Beach

Clinch River

Martinsville

Dismal Swamp

CANADA

ME
VT
NH
NY
MA
MI
CT RI
PA
IN
OH
NJ
WV
MD
DE
KY
Virginia
TN
NC
AL
GA
SC
FL

✪	Capital
•	City
~	River
—	State line

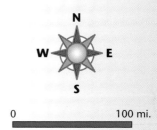

N
W   E
S

0   100 mi.

Glossary

adapt to change in a way that allows an animal or plant to live in new conditions

algae group of small plants that have no leaves, stems, or roots and grow in water or on wet surfaces

amphibian animal that lives in and around both water and land

anadromous fish that swim upstream from oceans to breed

aquatic living or growing in water

brackish mix of fresh and salt water

camouflage blending in with surroundings to hide

climate weather conditions that are usual for a place

crustacean animal with a tough outer shell that lives in water

ecosystem community of living things, together with the environment in which they live

endangered at risk of dying out

environment all the things that surround and affect a person, animal, or plant

erosion wearing away

estuary mouth of a river, where the saltwater tide flows in, and fresh and salt water mix

extinct no longer living

extirpated no longer living in a certain area

fall line place where multiple rivers fall from the uplands to the lowlands

filter to remove something while allowing something else to pass through

food web relationships of plants and animals within an ecosystem as prey and predators

forage to search for food

fossil remains or traces of a living thing from long ago that have turned to stone

fungus plant with no flowers, leaves, or green color

game wild animals hunted for food or sport

habitat place where a plant or animal lives

hibernate process by which an animal's body systems slow down for a period of time

Ice Age period of time when temperatures were lower and a large part of Earth was covered with ice

larvae young form of an insect, which looks like a worm and has no wings

living fossil plant or animal from an early period of Earth's natural history that is still living

mammal warm-blooded animal with a backbone; female mammals produce milk for feeding their young

marsh wet, low-lying area, often thick with tall grasses

migrate to move from one place to another for food or to breed

migratory species that moves from one place to another on a regular schedule

mollusk animal that has a soft body protected by a hard shell

mussel freshwater animal that has a soft body protected by a long, dark shell

native originally from a certain area

natural resource thing supplied by nature that is useful to human beings; coal, oil, air, water, and forests are natural resources

nocturnal active at night

pesticide substance such as DDT that is used to kill pests

plankton mass of tiny animals and plants that float together in the water

pollinate transfer material from one plant to another, enabling it to reproduce

predator living thing that hunts and eats other living things in order to survive

prey animal hunted for food by another animal

reintroduce bring something back to a place where it once was

resident person or animal that lives in a place

rodent any of a group of mammals with sharp front teeth used for gnawing

runoff rainwater that is not absorbed into the ground and drains into rivers and streams

SAV (submerged aquatic vegetation) plant that grows underwater

scavenger animal that feeds on rotting meat, dead things, or garbage

sediment dirt and other solid material that settles at the bottom of rivers and lakes

species group of plants or animals that look and behave the same way

suburb city or town just outside a larger city; *suburban* means having to do with a suburb

threatened group of animals whose numbers are decreasing, bringing the group close to endangerment

transplant move from one place to another

urban having to do with cities

wetland very wet, low-lying land

More Books to Read

Cerfolli, Fulvio. *The Animal Atlas.* Chicago: Raintree, 1998.

Hooper, Roseanne. *Life in the Coastlines.* Changassen, Minn.: Creative Publishing, 2000.

Kaner, Etta. *Animals at Work.* Tonawanda, N.Y.: Kids Can Press, 2001.

Spilsbury, Louise and Richard. *Plant Habitats.* Chicago: Heinemann Library, 2002.

Index

About the Author

Karla Smith grew up in a Navy family and moved several times before settling down in Suffolk, Virginia. She has been teaching third, fourth, and fifth graders social studies since 1969. When she is not teaching, Smith enjoys exploring Virginia's waters in a sailboat.